GHOSTS
FACT OR FICTION?

MY SCIENTIFIC THEORY.

Written by: SHEILA BER.

INTRODUCTION:

I majored in Science in 1991 from the University of Toronto.
Physics and Chemistry were especially, my favourite subjects.

I wrote this theory, as a result of my curiosity for the
supernatural phenomena.
My desire was to analyze the subject of ghosts' phenomenon,
from a perspective that is simple, and fundamentally scientific.

I hope that you'll find my Ghosts theory interesting.

Sheila Ber.

A SCIENTIFIC THEORY:

Ghosts? Ghosts I have never seen, nor did I ever believe they existed.

Ghost stories have been around for many centuries. The idea of apparitions from the spirit world goes back to the very beginnings of written history, and probably even farther back in oral traditions. A recent CBS News poll concluded that nearly half of all Americans believe in ghosts, and about 20 percent say they have seen or felt the presence of a ghost.

And yet mainstream science has long been clear and unequivocal: There is no scientific evidence of a supernatural explanation for ghost sightings. So how do we explain those incidents when rational people sincerely believe they have seen or felt a ghost?

What are some of the scientific, non-paranormal explanations for the phenomenon of ghost sightings?

While some accept ghosts as a reality, many others are skeptical of the existence of ghosts. For example, the vast majority of the scientific community believes that ghosts, as well as other supernatural and paranormal entities, do not exist. Skeptics often explain ghost sightings with the principle of Occam's razor, which argues that explanations should maximize parsimony with the rest of our knowledge.

They may suggest that, since few to none of us have ever had an interpersonal relationship with a ghost, but most or all of us have had an experience of self-delusion or have attributed a false cause to an event, that these options should be preferred in the absence of a great abundance of evidence.

They are also keen to note that most ghost sightings occur when our senses are impaired, and that the evidence is unreliable, because it doesn't occur when we have full use of our faculties.

Occasionally, the sincerity and motive of the claimant will be questioned. They might make up a haunting for a personal reason. For example, lingering of ghosts is typically associated with seeking justice or revenge. Ascribing such motives and powers to dead people could be interpreted as a scare tactic.

Human physiology may make us more susceptible to ghost sightings. Ghosts are often associated with a chilling sensation, but a natural animal response to fear is hair raising, which can be mistaken for chill. Also, the peripheral vision is very sensitive to motion, but does not contain much color or focused shapes. Any random motion outside the focused view can create a strong illusion of an eerie figure.

Some of the most popular haunted places are sacred burial grounds, cemeteries, sacred/power sites. Places where there are strong magnetic fields, places of religious worship, caves, places where acts of violence occurred, taverns, ancient ruins, castles, and perhaps even in one's home. Many people see ghosts of family members in their home.

After hearing many different stories and theories, I came up with my own rather simple, perhaps somewhat outlandish theory. A theory that may have scientific basis to it, and one day could also be proven.

The ghost stories that we hear are usually around Halloween time, in the Fall, at the end of October each year. So, I wondered why in October? Why mostly in the Fall (Autumn) time?

Well, Fall time is the rainiest time of the year. Rain water seeps deep into the ground, and saturates every living and non living objects. I assume therefore, that buried bodies underground, get soaked or re-hydrated with rain water, more so especially during Fall time.

Fully decomposed dead bodies, leave only the bones lying underground. Since the bones of the living have nerve cells. The bones of a dead body, I believe still probably retain their nerve cells, but they are in a dehydrated state. When they get hydrated however, by getting in contact with water seeping underground, they may have the capacity to function at some minimum level, which is of course at a different capacity, versus nerve cells in a living body.

To activate nerve cells, they require minerals, Oxygen, optimum temperature, and electricity. The first two variables are found in water, and of course the fourth one, electricity, is generated by the great cosmic electromagnetic and/or electrostatic force. When these variables are combined, they produce a synergistic action, which may spark life in dead bones, at an optimal temperature.

Since Electromagnetic fields have generally a profound effect on our nervous system, when we are alive, my understanding is, that they can also have an effect, at a different level, on nerve cells of a dead body lying underground, whether hydrated or dehydrated. The effect is greater however when a body, or even just the bones of a dead body, are in a re-hydrated state, again, due to contact with rain water seeping underground.

Electromagnetic and electrostatic fields exist nearly everywhere, and they gain more strength in saturated air, with a relative humidity that is high. The Fall season is relatively high in humidity. As we already know, electricity travels faster in water, because water is highly conductive.

Given the right conditions as mentioned above, they will serve as catalysts for electrochemical reactions in re-hydrated nerve cells inside dead bones, which may cause some level of renewed neural activity.

Every dead body has nerve cells of a different capacity, and therefore, will be affected accordingly, or perhaps will not be affected at all.

Those who have higher nerve cells capacity may, with the influence of external strong electromagnetic and electrostatic forces, have the ability to transmit electromagnetic signals or non verbal messages, from the underground and into the atmosphere.

The above forces may have the ability to spark a renewed neural activity in dead bones.

The signals may be interpreted in several forms. Some of the familiar ones are:

a) Non-verbal communication i.e. messages. b) Visible light faded shapes, formed by aura energy fields.

Again, electromagnetic force is perhaps capable of causing some level of awakening, in the re-hydrated nerve cells of dead bones.

Furthermore, nerve cells in the bones of a skull, have probably retained historical data of a dead person's past life. If energized and awakened by the factors as discussed above, and therefore have the capacity to release electromagnetic signals, such signals may be either weak or strong.

The signals from the dead, would probably be received and felt more easily, by the closest relative/s or friend/s of the dead person that's sending the signals. Relatives and friends may be the ones more familiar with such signals.

I asked myself again, Why ghosts stories around Halloween time? There must be another, better explanation.

The answer that I came up with is that it must be the Full Moon!

The sun and the moon exert forces and pressure on earth, and its objects. The most influential ones are the gravitational, electrostatic, and the electromagnetic force.

When the moon is in its full phase, it exerts stronger gravitational, electrostatic, and electromagnetic forces on earth, and everything on it and within it. These are combined forces of the moon, and of the sun, because the moon at the full phase, is situated directly between the Earth and Sun.

From the above scientific facts, I therefore deduced that the above forces which are relatively very strong during full moon phase period, may once again have an awakening or reviving effect, by generating a stronger energy field affecting not only the living, but the dead as well.

However, I must point out lunar and solar forces do not have sufficient power to bring the dead back to life, to form a fully functioning human body.

I'd like to conclude that when lunar and/or solar forces are exerted on dead bones, particularly during full moon phase, when these forces are at full strength, they might enable re-hydrated nerve cells inside the bones, to transmit electromagnetic signals that travel through the underground to the atmosphere. These signals weak or strong, may contain sensory information such as feelings of pain, joy, anger, or even revenge.

Again, these signals could be received, felt, and also be interpreted by the living entities, who may be more familiar with the content of such signals. Usually, it would be the relatives or friends of the dead person that transmit those signals.

*Two research groups reported in 1 issue of *Science* Journal, that bone marrow cells can travel to the brain and turn into nerve cells.

The bone marrow cells have the capacity to change into nerve cells. They appear to be making proteins that nerve cells make.
*The bone marrow does have a rich blood and nerve supply.

Feelings are nothing but electrical impulses felt by you in your brain. Feelings are usually often triggered by neuro-chemical agents.

Now, who says dead bones can't have feelings too?

*Similar to the evolution theory, unfortunately my theory would be very difficult to prove, and I accept that.

Nerve cells diagram:

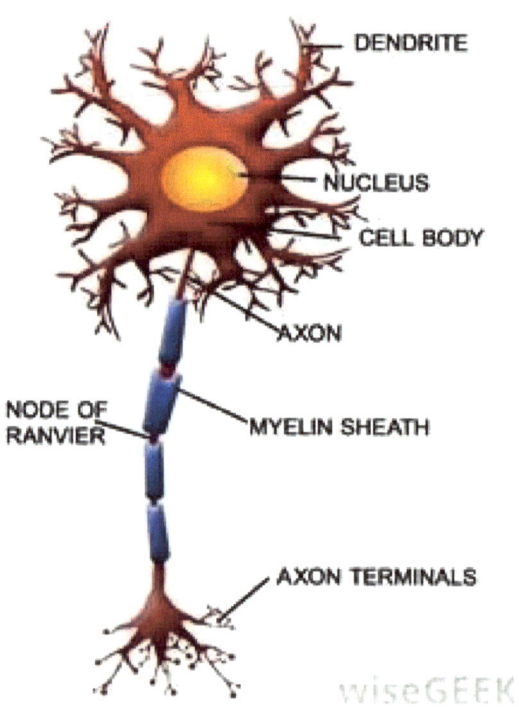

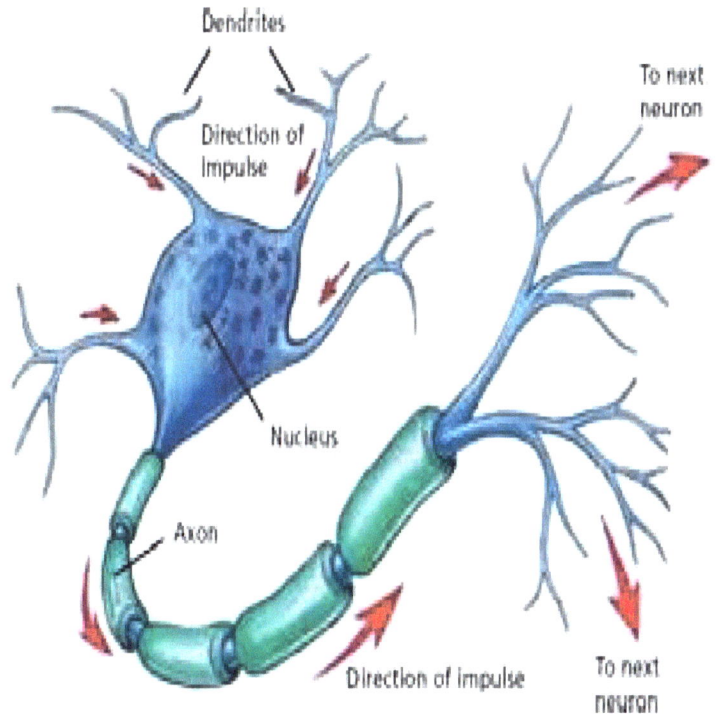

Keywords:

Dead bones, nerve cells, hydrated nerve cells, electromagnetic signals, electrostatic force, electromagnetic force, lunar and solar forces, spark of life, energy fields auras, neural activity, ghosts, spirits, supernatural.

References:

1) Foster, Russell G.; Roenneberg, Till (2008). "Human Responses to the Geophysical Daily, Annual and Lunar Cycles". Current Biology 18 (17): R784–R794. doi:10.1016/j.cub.2008.07.003. ISSN 0960-9822. PMID 18786384.
2) Carroll, Robert Todd (12 August 2011). "Full Moon and Lunar Effects". The Skeptic's Dictionary. Retrieved 22 October 2011.

Biography

I majored in Science in 1991 from the University of Toronto.
Physics and Chemistry were especially, my favourite subjects.

I worked in Microbiology and Chemistry, for about 12 years,
in the Pharmaceutical, cosmetics, and toiletry industries.
I was involved in Research & Development, analysis and in
formulations of large variety of products.

I am an unconventional person, at the same time I like things
straight, simple, and uncomplicated.

I have a tendency to analyze everything almost to death. Of course this habit has its positives but also some negative implications.
I like helping people. I view people, things, situations, from different perspectives, and try to remain neutral.

Our present digital world is a somewhat intimidating, but is rather promising at the same time. It is best to exercise the right balance in our lives.

SHEILA (SHULLA) BER

Disclaimer.

NEURAL ACTIVITY POTENTIAL IN DEAD BONES THEORY.

Now in: 1. www.createspace.com

2. www.amazon.com

3. www.kobobooks.com

www.ingramcontent.com/pod-product-compliance
Lightning Source LLC
Chambersburg PA
CBHW041620180526
45159CB00002BC/947